THE COUNTRY AT MY SHOULDER

THE COUNTRY AT MY SHOULDER

Moniza Alvi

Oxford New York
OXFORD UNIVERSITY PRESS
1993

Oxford University Press, Walton Street, Oxford OX2 6DP
Oxford New York Toronto
Delhi Bombay Calcutta Madras Karachi
Kuala Lumpur Singapore Hong Kong Tokyo
Nairobi Dar es Salaam Cape Town
Melbourne Auckland Madrid
and associated companies in
Berlin Ibadan

Oxford is a trade mark of Oxford University Press

First published in Oxford Poets
as an Oxford University Press paperback 1993

British Library Cataloguing in Publication Data
Data available

Library of Congress Cataloging in Publication Data
Alvi, Moniza.
The country at my shoulder / Moniza Alvi.
p. cm.—(Oxford poets)
1. Pakistan—Poetry.
I. Title. II. Series.
PR9540.9.A55C68 1993 821'.914—dc20 93—15598
ISBN 0-19-283125-9

1 3 5 7 9 10 8 6 4 2

Typeset by J&L Composition Ltd, Filey, North Yorkshire
Printed in Hong Kong

ACKNOWLEDGEMENTS

Acknowledgements are due to the editors of the following publications in which some of these poems first appeared: *Agenda*, *London Magazine*, *London Review of Books*, *Nutshell*, *Orbis*, *Poetry London Newsletter*, *Poetry Review*, *The Bound Spiral*, *The North*, *The Rialto*, *Transformation* (Rivelin Grapheme), *Writing Women*, and *Virago New Poets*.

A pamphlet collection of a number of these poems was co-winner of the 1991 Poetry Business Competition and was published in a joint collection *Peacock Luggage*, Alvi/Daniels (Smith/Doorstop Books, 1992).

I am grateful to my friends for their encouragement, and my parents for *Presents from Pakistan*.

CONTENTS

I Was Raised in a Glove Compartment

I was raised in a glove compartment.
The gloves held out limp fingers—

in the dark I touched them.
I bumped against the First Aid tin,

and rolled on notepads and maps.
I never saw my mother's face—

sometimes
her gloved hand would reach for me.

I existed in the quiet—I listened
for the sound of the engine.

Pilgrimage

On that dreamy late afternoon
the bushes alive with cabbage whites
Tom led me down the tangled pathway
between the mysteries of back gardens
and the top of the railway bank
where we were not supposed to play.

At last he pointed out a clearing
a sandy dip in the bank
and there it lay
a long fat turd.
It was Neil, Tom said.
Neil did it.
Other children had arrived
so we all gathered round
in a wide circle to study it
and it seemed to stare at us
burnished rich brown
almost kingly
as if the sun enjoyed it too—
an offering from a pale quiet boy.
Not one of us was disappointed.

Someone tried to burn it.
One by one we clambered up the bank
and trekked along the pathway.
Then later, at the tea-table
I thought of what Neil had left
in the vanishing sunlight.

Neighbourhood

Next door they were always fighting
calling each other Mr and Mrs
the names barking away
at the back of our chimney.

There were families with bitten
trickles of children
who pushed prams full of babies
junk and little dogs, smothered
and dressed in baby clothes.

On the run from my family troubles
I'd sneak to a back corner
of the playing field to savour
a terrible garden through the fence
where empty petfood cans were traps
in the Amazon grass.

And once Jim Skerry
bruised with dirt, never at school
came out through the din
behind his back door
wearing a hard brown plastic wig.
He sauntered off to the street.

Hydrangeas

The hydrangeas are massing
in gardens cherished by aunts.
Grimly ornamental, by tiled paths
they insist there is one garden—
theirs,
and if cats dare piss there
they swallow them
in huge mauve mouthfuls.

The hydrangeas are massing
heavy as cannonballs.
Guarding the house
they hiss to themselves
secretly.

They'll plan their own wedding—
invite no bride or groom.

The Afterworld

When your parents are gone
you will wear a white dress
and lock yourself inside
their empty house.

Daily, you'll pace the hollow boards
to the door, where you think you see
your father,
and the tree behind you
will be tapping greenly at the window.

A thousand times you'll write
at a child's desk
versions of the stories
they still pass to you in fragments.

Across them spreads your long hair—
it is still golden.
But you will not own anything
except the sudden sunlight
shining through your parents' hallway
up into your bedroom.

I Make Pencil Drawings, Scribbles, Bales of Hay

I make pencil drawings, scribbles, bales of hay,
bundles of twigs, a bonfire with parts torn out,
rough shapes of a man leaning on a woman

who leans on a child.

The distance is a simple line they turn to,
a tangle of rosebushes, the signature
of a dead queen—all black loops and angles.

I Would Like to be a Dot in a Painting by Miro

I would like to be a dot in a painting by Miro.

Barely distinguishable from other dots,
it's true, but quite uniquely placed.
And from my dark centre

I'd survey the beauty of the linescape
and wonder—would it be worthwhile
to roll myself towards the lemon stripe,

Centrally poised, and push my curves
against its edge, to get myself
a little extra attention?

But it's fine where I am.
I'll never make out what's going on
around me, and that's the joy of it.

The fact that I'm not a perfect circle
makes me more interesting in this world.
People will stare forever—

Even the most unemotional get excited.
So here I am, on the edge of animation,
a dream, a dance, a fantastic construction,

A child's adventure.
And nothing in this tawny sky
can get too close, or move too far away.

Hill

The hill heaps up her darknesses.

It is too cold for me to reach
the top. Yet she is always supreme,

dropping the lakes to her feet
and brushing them with her full skirt.

She'll never stoop down to her domain
now she has the luxury of a view.

In charge of greys and green-greys
she structures the late afternoon

with inky threads from stalling kites
and the hurtlings of solo footballers.

Slowly, she discards all detail—
offers me her dissolving horizons.

On Glastonbury Tor

The short grass at my feet is blown by a tornado.
Below me, fields level out to clouded glass.

St Michael's Tower—noble and lonely
as the last chess-piece on the board.
How I could hug a man like this!

Red Ridinghood's Plan

One iceberg December
the air mauve with cold
Red Ridinghood's mother
struck out her own heart.
She salted it, she seasoned it.

Her daughter swears to her
I'll never do that to mine
I'm going to visit Grandmother
for you haven't seen her
since the day you were born.

Red sets out across the forest.
The moon blackens like an old banana.
The trees have nothing to give but shadows.

She peers through her grandmother's window
net curtains half-way up.
The television blares and howls like a wolf.
Grandmother watches unseeingly
eyes swimming in absinth.

Red switches off the set
makes her grandmother and herself
cups of reviving tea.
She takes the cloth off her basket
and reassures the old lady

I've brought you my mother's heart.

The Garden

A plumtree swayed up to my window
stretching to pick me.
The path strode half-way up the grass
where it finished, straight and grey.

There were two grand flowerbeds—
the circular for sleeping,
the diamond for fighting
crammed with snap-dragons.

I dug up moss
between the paving-slabs
so that ants might flow—
until I stopped them.

The hollyhocks were the grown-ups
watching me from the heights,
shaking their rough, flat leaves.
They saw me running to the hedge,

Break up mushrooms sturdy as tables.
I asked my brother
to lie down by the lupins,
pretend to be dead.

Dream of Uncle

In that open grave by the villa
you rest on your side
in a well-tailored suit.
Your back still broad and comfortable—

Far more solid than smoke
that inters itself in the sky.

You're stranded so close to the house
I am loath to look at you.
Though you haven't diminished
you rise like a hill.

Your children still spin down the slopes.

Man of Tree and Fern

I must stand—
   my tree-legs are thin and stiff
my erection is bound
   it's all tree and fern
my arms are trapped in bracelets
   sharp as crescent moons.
Bundles of sticks
   grow from my shoulders.
Someone drew wings
   where I caught my breath.
I was a man of rank—
   you may tell me anything.
I can see with my eye-sockets
   dark as sunglasses.
I am more alive
   than the sad face carved
on my upper arm.
   I can bring you pets—
little bugs like pebbles
   the water runs over
and quick snakes that whirl
   like catherine wheels.

Before I burn

   the lamplight will flame
in tree-shapes
   table and chairs will blossom
for our feast.
   I'll stand by our lake-table
and you'll jump up

   to brush crumbs from your dress.

On Dunton Bridge

If I were an anxious child
I would wait by this bridge for
someone, anyone to cross it with me.

I daren't look down, my head
would bounce like a ping-pong ball,
nestle in the wasteland willow-herb.

I saw a couple fight here—Father
clutched a sobbing little boy,
Mother yelled *Get with me!*

at a girl in neat school uniform.
They all stood turned to stone
like Dunton Bridge. And I froze too.

Meeting an Ex-Pupil on a Spring Morning

Mary Jo, schoolgirl turned dental nurse
is smiling outside the surgery.
The wind clears the sky over Southwark Park.
In the sun-whitened street Mary Jo smiles—
a man in a poplar tree lops off branches,
leaves shine like silver on the fence.

Mary Jo smiles as if composing herself
for a photograph. This used to be my dentist.
Mary Jo knows. Tells me she's been looking
at my records. And then she smiles.
I see X-rays of my teeth blown up cinematic
on hoardings all over Bermondsey.

The blossoming trees brush against them
and hurl their petals to the ground
like sweetpapers. Mary Jo smiles.
A passing woman fills her baby's bottle
with Pepsi. And the sun, like flashlight,
bleaches the morning to the bone.

In Newspaper

My mother wrapped my cut-off plait in newspaper.
It might have waved its tail like a fish,
but it lay there, dignified.

My newspaper hair is robust and thickly plaited,
though sometimes a threat, as from a pitiless
comb, edges towards it.

Occasionally it holds a memory of snowflakes
from a first winter, settling on it,
crunchy as sugar—

Or a sense of fingers pulling it into its muscular
criss-cross, and within it the longing
to be wind-blown horse's hair.

At the bottom of the drawer it's a rope to some
nameless adventure, all secrets
tightly enmeshed.

Free from snipping, splitting, perming, it waits
long years to be touched in awe
and splashed with lamplight.

It's an ear of black corn, ripened in newspaper,
in my mother's room.

Throwing Out my Father's Dictionary

Words grow shoots in the bin
with the eggshells and rotting fruit.
It's years since the back fell off
to reveal paper edged with toffee-glue.
The preface is stained—a cloud rises
towards the use of the swung dash.

My father's signature is centre page,
arching letters underlined—I see him
rifling through his second language.

I retrieve it.
It smells of tarragon—my father's
dictionary, not quite finished with.

I have my own, weightier
with thousands of recent entries
arranged for me—like *chador*
and *sick building syndrome*
in the new wider pages.
I daren't inscribe my name.

PRESENTS FROM PAKISTAN

Indian Cooking

The bottom of the pan was a palette—
paprika, cayenne, dhania
haldi, heaped like powder-paints.

Melted ghee made lakes, golden rivers.
The keema frying, my mother waited
for the fat to bubble to the surface.

Friends brought silver-leaf.
I dropped it on khir—
special rice pudding for parties.

I tasted the landscape, customs
of my father's country—
its fever on biting a chilli.

Presents from my Aunts in Pakistan

They sent me a salwar kameez
peacock-blue,
and another
glistening like an orange split open,
embossed slippers, gold and black
points curling.
Candy-striped glass bangles
snapped, drew blood.
Like at school, fashions changed
in Pakistan—
the salwar bottoms were broad and stiff,
then narrow.
My aunts chose an apple-green sari,
silver-bordered
for my teens.

I tried each satin-silken top—
was alien in the sitting-room.
I could never be as lovely
as those clothes—
I longed
for denim and corduroy.
My costume clung to me
and I was aflame,
I couldn't rise up out of its fire,
half-English,
unlike Aunt Jamila.

I wanted my parents' camel-skin lamp—
switching it on in my bedroom,
to consider the cruelty
and the transformation
from camel to shade,
marvel at the colours
like stained glass.

My mother cherished her jewellery—
Indian gold, dangling, filigree.
But it was stolen from our car.
The presents were radiant in my wardrobe.
My aunts requested cardigans
from Marks and Spencers.

My salwar kameez
didn't impress the schoolfriend
who sat on my bed, asked to see
my weekend clothes.
But often I admired the mirror-work,
tried to glimpse myself
in the miniature
glass circles, recall the story
how the three of us
sailed to England.
Prickly heat had me screaming on the way.
I ended up in a cot
in my English grandmother's dining-room,
found myself alone,
playing with a tin boat.

I pictured my birthplace
from fifties' photographs.
When I was older
there was conflict, a fractured land
throbbing through newsprint.
Sometimes I saw Lahore—
my aunts in shaded rooms,
screened from male visitors,
sorting presents,
wrapping them in tissue.

Or there were beggars, sweeper-girls
and I was there—
of no fixed nationality,
staring through fretwork
at the Shalimar Gardens.

Arrival 1946

The boat docked in at Liverpool.
From the train Tariq stared
at an unbroken line of washing
from the North West to Euston.

These are strange people, he thought—
an Empire, and all this washing,
the underwear, the Englishman's garden.
It was Monday, and very sharp.

Luckbir

My Aunt Luckbir had full red lips,
sari borders broad like silver cities,
gold flock wallpaper in her sitting-room.
Purple curtains opened

on a small, square garden
where Uncle Anwar fed the birds
and photographed Aunt, her costume
draped over a kitchen stool,

the backdrop—a garden fence and roses.
Luckbir found her Cardiff neighbours
very kind—thanked them in a letter
to *Woman's Own*,

spoke to me warmly of Jane Austen
remembered from an overseas degree.
Aunt had no wish to go out, take a job,
an evening class.

Picking at rice on pyrex
she grew thin—thinner.
In my dreams she was robust,
had a Western hairstyle, stepped outside.

After she died young
Uncle tried everything—
astronomy, yoga, cookery.
His giant TV set flickers into life—

a video of the ice-skating championship.
Has he kept Aunt's clothes,
let their shimmer slip through his hands?

The Country at my Shoulder

There's a country at my shoulder,
growing larger—soon it will burst,
rivers will spill out, run down my chest.

My cousin Azam wants visitors to play
ludo with him all the time.
He learns English in a class of seventy.

And I must stand to attention
with the country at my shoulder.
There's an execution in the square—

The women's dupattas are wet with tears.
The offices have closed
for the white-hot afternoon.

But the women stone-breakers chip away
at boulders, dirt on their bright hems.
They await the men and the trucks.

I try to shake the dust from the country,
smooth it with my hands.
I watch Indian films—

Everyone is very unhappy,
or very happy,
dancing garlanded through parks.

I hear of bribery, family quarrels,
travellers' tales—the stars
are so low you think you can touch them.

Uncle Aqbar drives down the mountain
to arrange his daughter's marriage.
She's studying Christina Rossetti.

When the country bursts, we'll meet.
Uncle Kamil shot a tiger,
it hung over the wardrobe, its jaws

Fixed in a roar—I wanted to hide
its head in a towel.
The country has become my body—

I can't break bits off.
The men go home in loose cotton clothes.
In the square there are those who beg—

And those who beg for mercy.
Azam passes the sweetshop,
names the sugar monuments Taj Mahal.

I water the country with English rain,
cover it with English words.
Soon it will burst, or fall like a meteor.

The Sari

Inside my mother
I peered through a glass porthole.
The world beyond was hot and brown.

They were all looking in on me—
Father, Grandmother,
the cook's boy, the sweeper-girl,
the bullock with the sharp
shoulderblades,
the local politicians.

My English grandmother
took a telescope
and gazed across continents.

All the people unravelled a sari.
It stretched from Lahore to Hyderabad,
wavered across the Arabian Sea,
shot through with stars,
fluttering with sparrows and quails.
They threaded it with roads,
undulations of land.

Eventually
they wrapped and wrapped me in it
whispering *Your body is your country.*

Map of India

If I stare at the country long enough
I can prise it off the paper,
lift it like a flap of skin.

Sometimes it's an advent calendar—
each city has a window
which I leave open
a little wider each time.

India is manageable—smaller than
my hand, the Mahanadi River
thinner than my lifeline.

When Jaswinder Lets Loose her Hair

When Jaswinder lets loose her hair
it flows like a stream
that could run around a mosque.

Below her knees it thins
to a wild sparseness
as if the wind tugged at it

when as a girl she sped
across the formal garden.
I crave Jaswinder's hair,

coiled—dark and smooth
as a glazed pot, falling—
a torrent of midnight rain.

The Asian Fashion Show

It's a charity show—a diversion
from chat, chips and baked beans.

I usher Asian girls towards the few
rows of chairs in the school hall.

College students parade—young women
drift across the dusty parquet,
modelling informal clothes and saris
for special occasions, salwar kameez
so stunning they ward off insults,
silks that could brush against years
of criss-cross graffiti.

Afterwards
Ruskhana shows me her Fashion Folder—
Asian clothes, pin-ups,
photographs of Imran Khan.
She tells me where to buy such outfits—
the best shops in Wembley and Moorgate,
asks if I have relatives in Pakistan.

By morning she's found out from her dad
the price of my air-ticket to Islamabad.

The Draught

I

In winter
there'll be draughts,
my mother warns.

Evenings will be chilly—
I'll huddle by the fire
which gives off fumes.
My aunts wear shawls
over salwar kameez and
socks with pretty sandals.

I'll travel with layers of clothes,
the recommended thermal underwear,
and I'll look up at those huge stars.
A road winds back inside me
like the Karakoram Highway
through mountains, gorges.

What to expect?
Aunt Shazia waits on Uncle.
Her dupatta dips
as she bends and offers sweetmeats.
Uncle Ali is building a new home.
My aunts don't know about it yet—
they'll live there soon.

There'll be draughts—
everyone leaves doors open.
Gusts of stories.
Bandits in the foothills
held up Amir's car.
They took everything.
Someone died on the journey.

II

In Shah Jehan's garden
where children splash their toes
in pools and channels
I'll hear an echo of swollen waters,
the thunder of warring countries,
and I'll wear my trousers
which mustn't be too tight,

envy the trees—almond, mango,
citrus, mulberry—riotous
in the formal gardens.
I'll shut my eyes
so I can't be seen
in the towns of invisible women.
New words perch on my tongue.

In my dreams I trawl
towards a brilliant sun
with a burden of gold jewellery,
push down traffic-clogged streets.
A trap of bones
is set around my neck.

And the draught?
The great draught
blowing me
to my birthplace.
Will it sweep the colours
off the salwars,

the smile on and off my face
as I try to hold a house
of Eastern women,
turning it in my hands
in the Shish Mahal,
palace of tiny mirrors.

Domain

Within me lies a stone
like the one that tries

to fill the mango.
Inside it is the essence

of another continent.
I fear its removal—

but how much better
it could be

to take it in my arms
and race away with it!

Etching

for Javaid

All the sparrows in the black and white miniature
are golden. Other small bodies are eels,
locusts, zig-zags and crowbars.
A pale wool shadow of a scarf is suspended
in the teeming air.
The boy, who might be a girl,
wears a turban, and has two black eyes
from centripetal pressure.
He knows he can draw elegance
into the tiny white cracks in his eyes.
He doesn't know what suffuses him
where chaos and enchantment
jostle like relatives.

He is in a garden
He is underwater
He has one arm

The Air Was Full of Starfish

She was born
with the fervour of a bible-class
a senator with his whisky glass.

She broke, she entered.

Her mother bathed her—
a bowl on a check table-cloth
bracelets chinking
scissors and knives on the wall.

Mother and Father sat in parallel armchairs.
Her sisters were flinging themselves
down on mattresses.

The air was full of starfish
the air was full of flying machines.
Her dress wafted out like a powder puff.

Eating pudding in the marbled pavilion
private as prayer
she was changing minute by minute
like the amber moon stuck up in the sky.

The child—the dreamer, the dream
running in a torn camisole.

Her footprints stay where she left them.

The Great Pudding

I would like to announce this great pudding,
the biggest yet, blended by mixer-lorry,
steamed in a stainless-steel vat.
The Prince has ordered his portion,
and the giant boiling dish—
once used to soften willow-bark,
is now on show.

Children roll the pudding
down the street.
It's so solid
it has to be broken up
with pickaxes.

The Mayor takes a crack at it.
Outside the town hall
the pudding spins
like a globe,
blocking the square.

Coins glint like fixed stars—
it's hard to prise them out.
It isn't burnt, or overcooked.
Everyone aims for the core—
the dark, moist heart
where fruit hugs fruit.

When We Ask to Leave Our City

They say

Why don't you instead
cut off your hair
tamper with the roots?

So we ask to leave the Earth forever.

They say *What! Leave the Earth forever?*
And they offer a kind of umbilical cord
to connect us up to the coldest stars.

We'll stay in our city
opt for decent bread
lean our unruly heads
against the wall.

On Finding a Letter to Mrs Vickers on the Pennine Way

A bird with a torn tail hops under ferns
and points its beak to the wall.

A letter to Mrs Vickers is trodden into the path—
colours have run into edges soft as cotton.

Mrs Vickers, Mrs Vickers
you have won, you have almost won
a Ford Escort. We of the Prizes Department
are sending you a draft of the Award Certificate.

Earth trickles over it like a child's pattern.

Mrs Vickers, calling your number at Stoneway
we would like to tell you
you're in with a winning chance.
Don't miss the cellophane window.

It shines like a dirty film of ice.

Mrs Vickers, don't forget to tell us
all about yourself.
Then tread this well into the path
where the mossy fronds dart like fishes—

And the bird fans out its broken tail.

Housebreaker

I dash my fist against the white walls—
they dent coldly like ice-cream.
Smug as a conjurer in the quiet hallway,
my face grazes slightly in the light.
Busy in the bedroom, I'm spell-blinding,
cupping nightmares in my hands,
turning them into silk scarves.
I long to feel the charred paper breath
of sleepers rising and falling on my skin.
If they wake, I'll loom up in monochrome,
I'll tell them how to catch a rat—
dye an ear of barley blue, for bait.

Moonlight whitens the lawn. A drained
swimming pool. Car on the track.

Is that a guitar in the corner of the room?
I could tell them how the lute is played differently.
Downstairs I'm a strange bird in a classical setting.
I fly into china, make cool, surgical music,
slide on the rugs, skating for information.

The house I gut goes down like a ship in the night.
At dawn, like a conspiracy, it rises behind my back.

Afternoon at the Cinema

You are watching a film you can't understand
with the man you once lived with
who reassures you that he can't understand it either
but he supposes it's not supposed to be exactly—
What was the word he used?

The film—you've seen it before—
it was a mystery then, and now you've missed
the sheet with interpretations on it.
Perhaps you'll get one later at the door.
The screen pours out its pictures
mirrors swaying like water
mirrors shivering in fragments
to walk through, turn back through
while pairs of lovers interchange and
there are murders and crowds and everyone
is tumbling off the screen.

You are watching a film you can't understand
with the man you once lived with.
You're in flames and falling on a mirror.
Was that something that he said?

Pink Ear-Plugs

She found one of Beth's tiny pink ear-plugs
in the bed with the grit and crumbs
sometime after she'd left.

And then she imagined her
in grander beds, a quilt
falling away to expose her—
a featherless bird.
Sometimes she saw her furred—
a leopard covered in rosettes
she'd slip from the bed,
leap up a tree.

Or she glimpsed her lying with someone—
a hand resting on a hand
to see who has the smallest palm,
the longest fingers.

You Are Turning Me into a Novel

I hear your pen sliding on paper—
my skin thins like a page,
my eyelids quaver like words.

In the great silent hour
you are giving me a title,
fashioning me, coaxing me.

When I ask myself Who am I?
there'll be readers
attentive

on cool afternoons
bending back my spine.
I am the Knave of Dreams

rising from the margins.
I hope you don't think
it is my life

as you hole-punch the pages,
finger the ends of lines.
You are travelling towards me

at the speed of an idea.
The words you posted
are inside my mouth.

You are giving me a white cover.
Somewhere in each chapter
I do something terrible.

The Fire-Walk

I fire-walked for you
at the World of Adventures—
except I was unprepared
no five-hour initiation
to develop my resistance.
I dithered a little
removing my shoes—
and then I was off
taking brisk steps.
But the coals were not
the right kind
or they were not
at the right temperature.
My feet were flat—
I made too much contact
with the flames.
I stood still, transfixed
on the coals
and that was fatal.
More than once I stumbled
but was impelled
to continue walking.
I was not like those
who are empowered
by a fire-walk
who lose half a stone
in a fortnight
or start a new business.
My skin began to melt.

No one was held liable—
not even you.
You gave off stars
like a sparkler.
Fire was your element.

The Bees

Her breast is a face that looks at him.
On a quiet night he's convinced he can
open it and riches will tumble out—
pearls, golden brooches, honey.

The bees swarm into his mouth,
settle behind his ears, buzzing.
The breast is a safe—
But there's no getting in.

Ladybirds, Icebirds

Show me the ladybirds,
their colonies glistening
like a bloodstream
over the rocks and snow
high in the mountains.
Let me see the plant whose
lid of ice prevents the sun
from drying up its centre.
There's snow stained red
from minute cells of algae.

Penguins protect their eggs
from freezing, desire
a chick so strongly they'll
incubate a piece of ice.
I could take an icebird—
beak, breast, feathers
in my gloved hand,
now my hair is whitening
like the arctic fox
in winter.

Spring on the Hillside

You coax the flames, choosing
the logs thoughtfully.
You're happy
as if a bright ghost of yourself
moves through the openings
in your wigwam of sticks.

We stumble down the valley
to make phonecalls.
The evening sun
wanders like torchlight
across the trees
the agate river.

I have never known wild flowers
live so long in a vase
as in the hills of Llandogo
where the sleet falls like blossom,
the stone path trails upwards
further than we know.

The Bed

We have travelled many miles to find this bed,
scanned tedious columns of small print,
waited in queues at bus stops.
And now we think we have it—

But we'll have to get it home—
no one can deliver it for us.

We must test it.

Is it wide enough for a family?
Will it hold the tempests of our dreams?

And when we are accustomed to it,
when the pillows burst their stomachs
and ecstatic feathers fly towards the cornice,
then perhaps, we'll have that river
in the middle of the bed—

Where in the ancient song
the King's horses could all drink together.

OXFORD POETS

Fleur Adcock
Moniza Alvi
Kamau Brathwaite
Joseph Brodsky
Basil Bunting
Daniela Crăsnaru
W. H. Davies
Michael Donaghy
Keith Douglas
D. J. Enright
Roy Fisher
Ivor Gurney
David Harsent
Gwen Harwood
Anthony Hecht
Zbigniew Herbert
Thomas Kinsella
Brad Leithauser
Derek Mahon
Jamie McKendrick
Sean O'Brien
Peter Porter
Craig Raine
Henry Reed
Christopher Reid
Stephen Romer
Carole Satyamurti
Peter Scupham
Jo Shapcott
Penelope Shuttle
Anne Stevenson
George Szirtes
Grete Tartler
Edward Thomas
Charles Tomlinson
Marina Tsvetaeva
Chris Wallace-Crabbe
Hugo Williams